CONSENSUS ON OPERATING PRACTICES FOR THE SAMPLING AND MONITORING OF FEEDWATER AND BOILER WATER CHEMISTRY IN MODERN INDUSTRIAL BOILERS

AN ASME RESEARCH REPORT

prepared by the
Sampling and Monitoring Task Group and the Water Technology
Subcommittee of the Research and Technology Committee on
Water and Steam in Thermal Systems of the
American Society of Mechanical Engineers.

ASME, Three Park Avenue, New York, NY 10016, USA

CONSENSUS ON OPERATING PRACTICES FOR THE SAMPLING AND MONITORING OF FEEDWATER AND BOILER WATER CHEMISTRY IN MODERN INDUSTRIAL BOILERS

❒ PREFACE ❒

The Water Technology Subcommittee of the ASME Research and Technology Committee on Water and Steam in Thermal Systems has established a "Consensus on Operating Practices for the Sampling and Monitoring of Feedwater and Boiler Water Chemistry in Modern Industrial Boilers". This publication provides recommendations for water chemistry monitoring and is a companion to the ASME "Consensus on Operating Practices for the Control of Feedwater and Boiler Water Chemistry in Modern Industrial Boilers," "A Practical Guide to Avoiding Steam Purity Problems in the Industrial Plant," and "Consensus on Operating Practices for Control of Water and Steam Chemistry in Combined Cycle and Cogeneration Power Plants". These documents provide guidance for avoiding the penalties of severe corrosion or deposition in steam generation systems and end user equipment.

Individuals associated in the compilation and preparation of this document consisted of manufacturers, operators and consultants involved in all aspects of the design and operation of industrial and utility boilers, heat recovery steam generators, boilers and the associated equipment in the full power train. The members of the task group, committee and others making significant contributions to this document are listed in the Acknowledgement section.

The information in this document will be reviewed by the Research and Technology Committee and revised and reissued as necessary to comply with advances in boiler design or water treatment technology.

It is my pleasure to thank all of those individuals who contributed considerable time, and freely shared their knowledge and experience, to the preparation and publication of this consensus.

Robert T. Holloway
Chair, ASME Research and Technology
Committee on Water and Steam in
Thermal Systems

◻ ACKNOWLEDGEMENTS ◻

This document was prepared by the Sampling and Monitoring Task Group and the Water Technology Subcommittee of the Research and Technology Committee on Water and Steam in Thermal Systems of the American Society of Mechanical Engineers. Recognition is hereby given to the following members of these groups for their contributions in the preparation of this document.

I. J. Cotton — Co Chair
R. W. Light — Co Chair

A. Banweg
T. Beardwood
D. M. Bloom
I. J. Cotton
R. J. Cunningham
D. Daniels
D. B. Dewitt-Dick
S. B. Dilcer, Jr.
J. C. Dromgoole
R. T. Holloway
J. Isaac
M. Janick

C. M. Kulick
R. W. Light
L. M. Machemer III
W. Moore
L. Olavessen
J. O. Robinson
J. Sabourin
K. A. Selby
D. E. Simon II
K. Sinha
T. J. Tvedt, Jr.
D. K. Woodman

❏ CONTENTS ❏

1	❏	Introduction	1
2	❏	Scope	3
3	❏	Importance of Effective Monitoring	5
4	❏	Discussion of Parameters to be Monitored	7
5	❏	Sampling	9
6	❏	Analytical Testing	39
7	❏	Test Methods	41

Tables

1	❏	Guidelines for Sampling Frequency by Location and Test Sodium Zeolite Make-up	13
2	❏	Guidelines for Sampling Frequency by Location and Test High Purity Make-up	15
3	❏	Purge Times Required for Representative Sampling of Water	18
4	❏	Test Methods	19

Figures

1	❏	Recommended Sample Point Locations Softened Water Make-up	10
2	❏	Recommended Sample Point Locations High Purity Make-up	10
3	❏	Sampling Nozzle Installation	11

Bibliography   43

□ SECTION 1 □
INTRODUCTION

All units and sub-systems in the power train of industrial, utility and heat recovery steam generation facilities must be adequately sampled and monitored to maintain operation as desired and to prevent damage to individual pieces of equipment. Besides the obvious effects of failure to maintain design production of treated water and steam on plant production costs, failure to adequately monitor operating conditions can cause failure of major system components and cause unplanned outages.

This document has been prepared by the Water Technology Subcommittee of the ASME Research and Technology Committee on Water and Steam in Thermal Systems. This publication provides recommendations for water chemistry monitoring and is a companion to the ASME "Consensus on Operating Practices for the Control of Feedwater and Boiler Water Chemistry in Modern Industrial Boilers," "A Practical Guide to Avoiding Steam Purity Problems in the Industrial Plant," and "Consensus on Operating Practices for Control of Water and Steam Chemistry in Combined Cycle and Cogeneration Power Plants". These documents provide guidance for avoiding the penalties of severe corrosion or deposition in steam generation systems and their auxiliary steam user equipment.

The information in this document will be reviewed by the Research and Technology Committee on a regular basis and revised and reissued as necessary to comply with advances in boiler design or water treatment technology.

The reader is urged to read all of the associated text to fully understand the impact of the recommendations contained in this document.

☐ SECTION 2 ☐
SCOPE

This document provides recommendations for the sampling and analytical testing required to maintain the suggested water chemistry limits outlined in the "Consensus on Operating Practices for the Control of Feedwater and Boiler Water Chemistry in Modern Industrial Boilers" and the "Consensus on Operating Practices for Control of Water and Steam Chemistry in Combined Cycle and Cogeneration Power Plants."

The boilers covered by these recommendations are:

- industrial water tube, high duty, primary fuel fired, drum type with superheaters and/or process restrictions on steam purity
- industrial water tube, high duty, primary fuel fired, drum type without superheaters
- industrial fire tube, high duty, primary fuel fired
- industrial coil type, water tube, high duty, primary fuel fired
- marine propulsion, water tube, oil fired, drum type
- electrode type, high voltage, recirculating jet type
- single pressure industrial heat recovery steam generators

Many of these sampling and monitoring recommendations may also be applicable to other types of steam generating systems, but good engineering judgment must be used before they are applied to:

- multiple pressure heat recovery steam generators
- mobile locomotive boilers
- boilers of copper or other unusual materials of construction
- immersion type, electric boilers, and low voltage electrode type boilers
- heating boilers of special construction
- waste heat boilers of unusual design
- hot water boilers
- oil field steam flood steam generators

The monitoring of specific types of make-up pretreatment, condensate treatment, and internal chemical treatment is outside the scope of this document. However, the requirement for such monitoring is clearly implied by the suggested values for feedwater quality in ASME Vol. CRTD 34.

☐ Section 3 ☐
IMPORTANCE OF EFFECTIVE MONITORING

Effective monitoring of makeup water, condensate, feedwater and boiler water qualities and steam purity is necessary to control deposition and corrosion within the boiler, steam, and condensate systems. The absence of adequate monitoring and control can lead to operational upsets or unscheduled outages and is ill advised from the point of view of safety, economy and reliability.

A key component of effective monitoring is sampling. The sample must be truly representative of the stream being monitored. This requires a proper sampling system, adequate system flushing, the required sample velocity, and appropriate sample handling during the time between sample collection and analysis. If proper procedures are not followed, analytical results may not be representative of the stream being monitored and the results will be meaningless and potentially harmful if used as the basis of operating decisions.

❐ Section 4 ❐
DISCUSSION OF PARAMETERS
TO BE MONITORED

❐ 4.1 Steam Purity

The steam purity required for any given boiler system is dictated by the use of the steam. Steam of inadequate purity can cause deposition and corrosion in superheaters, steam lines, control valves and turbines. Steam contaminants can also adversely affect catalyst performance and product purity. The ASME "Consensus on Operating Practices for the Control of Feedwater and Boiler Water Chemistry in Modern Industrial Boilers" and "Consensus on Operating Practices for Control of Water and Steam Chemistry in Combined Cycle and Cogeneration Power Plants" provide steam purity targets for typical industrial uses of steam.

The purity of steam is usually monitored by measuring the cation conductivity or sodium content of condensed steam samples. In special situations, as noted in the Combined Cycle and Cogeneration document, silica and/or cation conductivity will be measured.

Since producing steam of the required purity is of utmost importance in most systems, water chemistry parameters such as silica, dissolved solids and alkalinity must be controlled so that steam purity requirements are met. When problems are encountered in producing steam of the required purity, it is recommended that the user consult ASME publication "A Practical Guide to Avoiding Steam Purity Problems in the Industrial Plant".

❐ 4.2 Dissolved Oxygen

Feedwater dissolved oxygen monitoring is essential to identify any deaerator malfunction or points of oxygen ingress. Improper concentration of dissolved oxygen in the feedwater can lead to corrosion and failure of feedwater heaters, economizers and the entire boiler system.

❐ 4.3 Iron, Copper and Hardness

Feedwater iron, copper and hardness can all lead to the formation of insulating deposits on tube surfaces. If excessive deposits develop, tube overheating and failure can occur. In addition, porous deposits provide sites for the local concentration of soluble species, which may cause corrosion. Consequently, it is important to effectively monitor the level of these contaminants in the boiler feedwater so boiler reliability can be maintained.

❏ 4.4 pH

Effective pH control is necessary to control corrosion, metals transport and deposition throughout the boiler system.

❏ 4.5 Organic Matter

There are numerous types of organic matter that can contaminate boiler feedwater. Humic acids, tannins, lignins, polysaccharides, sugars and oily matter are among the most common. Many organic contaminants contribute to boiler water foaming as well as feedwater heater and boiler deposits. Decomposition products often volatilize into the steam increasing the cation conductivity. It is usually necessary to identify the likely feedwater organic contaminants before an effective monitoring program can be established so that any necessary corrective action can be implemented.

❏ 4.6 Silica

Control of silica is necessary to avoid excessive levels in the steam due to volatilization. Excessive silica can lead to deposition on turbine blades and other steam utilizing equipment. Where silica volatilization is not a consideration, it is still important to control silica to avoid formation of insulating deposits on the steam generator tubes.

❏ 4.7 Total Alkalinity

Control of alkalinity is necessary to ensure corrosion inhibition and to minimize boiler water carryover tendency. Alkalinity levels are also necessary to ensure silica solubility and proper functioning of other corrosion and deposit control treatment chemicals.

❏ 4.8 Conductivity and Total Dissolved Solids

Conductivity and total dissolved solids are important to minimize potential for carryover from boiler systems in order to maintain steam purity. Too high a level of conductivity or total dissolved solids can also result in increased potential for deposition and corrosion in boiler systems.

❐ Section 5 ❐
SAMPLING

❐ 5.1 Sampling — General

Successful control of boiler system water chemistry requires that meaningful and representative samples be acquired at various points throughout the boiler system. Samples and/or sample points must meet the specific requirements (e.g. flow, temperature, velocity) of the individual tests or the continuous analyzers that utilize them. They must also allow the user to acquire samples in a safe manner; e.g. provide cooling as necessary, route the sample to a protected area, etc.

❐ 5.2 Sample Tap Location

Recommended sample points for make-up, feedwater, blowdown, steam, and condensate are shown in Figures 1 and 2. Locations are included for both grab samples and continuous analyzers. Systems operating with high purity make-up typically have additional and more stringent sampling requirements.

❐ 5.3 Sample System Design and Conditioning

Typically, samples that exceed 25°C [77°F] must be cooled prior to collection to maintain a representative sample, prevent loss of sample volatiles and to assure the safety of collection personnel. Temperature should ideally be maintained at a constant 21-35°C [70-95°F], with 25°C [77°F] being most preferred. Certain continuous in-line analyzers, however, are capable of accepting higher temperature samples. Temperature tolerance varies from instrument to instrument, so the manufacturers recommendations or requirements should be checked prior to designing a sample conditioning system for any new continuous analyzers. Temperature compensated analyzers may not fully correct for the effect of sample temperature on analyzer readings. Therefore, it is recommended that secondary coolers be used to achieve a proper sample temperature.

When multiple streams are combined or chemicals injected, samples should be collected sufficiently downstream to assure good mixing. This is typically a minimum of 25 pipe diameters for turbulent flow or 50 pipe diameters for laminar flow from the blending or injection point.

In addition, several other general requirements exist for sample conditioning systems. These requirements serve to minimize the pickup or loss

Recommended Sample Point Locations

Refer to Table 1 for Details.

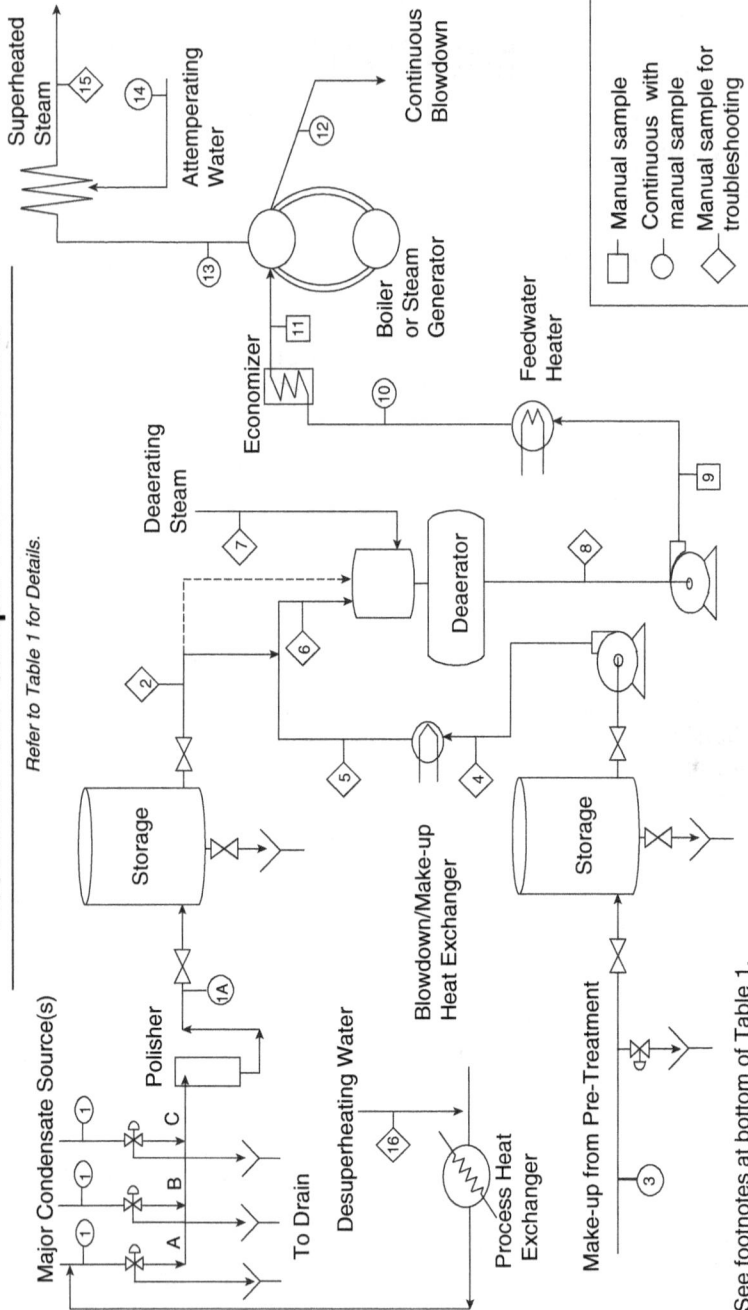

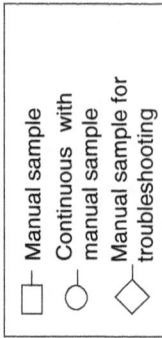

Figure 1 Soft Water Make-Up

See footnotes at bottom of Table 1.

Legend:
- ☐ Manual sample
- ○ Continuous with manual sample
- ◇ Manual sample for troubleshooting

Table 1 - Guidelines for Sampling Frequency by Location and Test for Softened Make-up (1). Refer to the Figure 1 for Sample Locations.

Sample Point No.	Sample Description	Parameter	Continuous Analyzer (2)	Minimum Recommended Test Frequency for Manual Testing	Remarks
1, 1A	Major condensate sample points	pH	X	Once/shift	
		Total hardness		Troubleshooting	
		Total iron		Once/day	
		Total copper		Troubleshooting	
		Total organic carbon	X	Troubleshooting	Continuous TOC is required if pH and conductivity do not pick up contaminant process leaks.
		Silica		Troubleshooting	
		Specific conductivity	X	Once/day	If manual testing is used as back up to conductivity dump, a greater frequency is needed.
2	After condensate storage	Same as for sample point 1, 1A		Troubleshooting	
3	Treated make-up water before storage	Total hardness	X	Once/shift	Increase manual frequency to once every two hours if analyzer is off line.
4, 5 or 6	Treated make-up after storage, preheater, or condensate addition	Same as for sample point 3		Troubleshooting	
7	Deaerating steam	Same as for sample point 13		Troubleshooting	
8	Deaerator storage outlet	Same as for sample point 10		Troubleshooting	

(*Continued*)

Table 1 - (Continued)

Sample Point No.	Sample Description	Parameter	Continuous Analyzer (2)	Minimum Recommended Test Frequency for Manual Testing	Remarks
9 or 10	Feedwater at economizer inlet (DA effluent)	Total hardness	X	Once/shift	
		Silica		Troubleshooting	
		Specific conductivity		Troubleshooting	
		Total iron		Once/shift	
		Oxygen	X	Once/day	
		pH		Once/shift	
		Copper		Troubleshooting	
11	Feedwater after economizer	Total iron		Once/shift	
		Copper		Troubleshooting	
12	Continuous blowdown	Silica	X	Once/shift	
		Specific conductivity	X	Once/shift	
		OH Alkalinity	X (pH)	Once/shift	
		M Alkalinity		Troubleshooting	
		Product test		Once/shift	To be specified by water treatment company.
		pH		Troubleshooting	
13	Saturated steam	Sodium (3)	X (4)	Troubleshooting	
		Silica		Troubleshooting	
		Specific conductivity (3)	X	Once/shift	

Table 1 - (Continued)

Sample Point No.	Sample Description	Parameter	Continuous Analyzer (2)	Minimum Recommended Test Frequency for Manual Testing	Remarks
14	Attemperation water	Sodium		Troubleshooting	
		Silica		Troubleshooting	
		Specific conductivity	X	Once/shift	
15	Superheated steam	Same as for sample point 13		Troubleshooting	
16	Desuperheat water	Specific conductivity		Once/shift	

(1) Additional tests might be required for higher-pressure boilers or steam generators.

(2) Continuous analyzers should be compared to grab samples at least daily. Increase manual frequency to once/2 hours if analyzer is off line.

(3) Sample point location and sample system are critical.

(4) If required by steam use.

Recommended Sample Point Locations

Refer to Table 2 for Details.

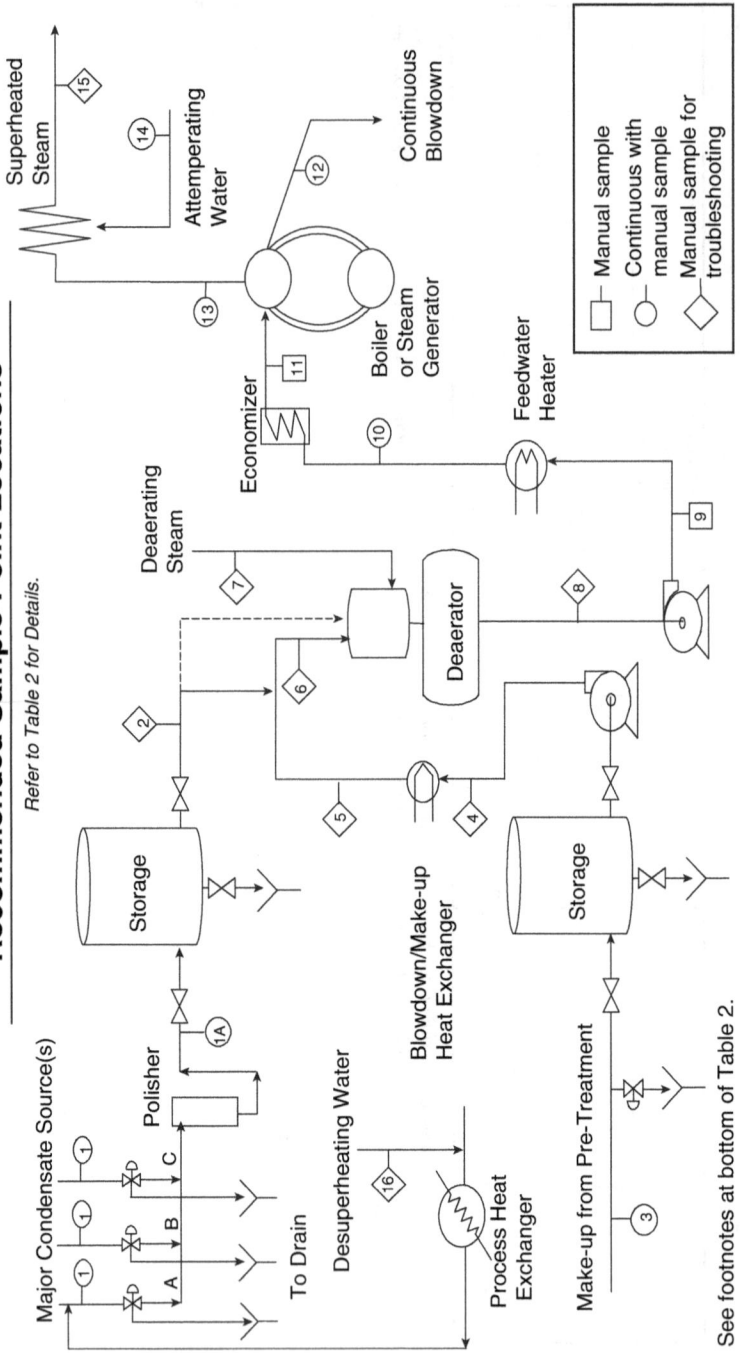

Manual sample

Continuous with manual sample

Manual sample for troubleshooting

See footnotes at bottom of Table 2.

☐ 14

Table 2 - Guidelines for Sampling Frequency by Location and Test for High Purity Make-up (Typically <10 μS) (1). Refer to the Figure 2 for Sample Locations.

Sample Point No.	Sample Number Description	Parameter	Continuous Analyzer (2)	Minimum Recommended Test Frequency for Manual Testing	Remarks
1, 1A	Major condensate sample points	pH	X	Once/shift	
		Total hardness		Troubleshooting	
		Total iron		Once/shift	
		Millipore iron		Once/shift	Not required if total iron is determined.
		Total copper		Troubleshooting	
		Total organic carbon	X	Troubleshooting	Continuous TOC is required if pH and conductivity do not pick up conta- minant process leaks.
		Silica		Troubleshooting	Once/shift if no cation conductivity
		Specific conductivity	X	Once/day	Not required if cation conductivity is run.
		Cation conductivity	X		
2	After condensate storage	Same as for sample point 1, 1A		Troubleshooting	
3	Treated water before storage	Total hardness		Troubleshooting	
		Specific conductivity	X	Troubleshooting	
		pH	X	Troubleshooting	
		Sodium		Troubleshooting	
		Silica	X	Troubleshooting	

(Continued)

Table 2 - (Continued)

Sample Point No.	Sample Number Description	Parameter	Continuous Analyzer (2)	Minimum Recommended Test Frequency for Manual Testing	Remarks
4, 5 or 6	Treated make-up after storage, preheater, or condensate addition	Same as for sample point 3		Troubleshooting	
7	Deaerating steam	Same as for sample point 13		Troubleshooting	
8	Deaerator storage outlet	Same as for sample point 10		Troubleshooting	
9 or 10	Feedwater at economizer inlet (DA effluent)	Total hardness		Troubleshooting	
		Silica		Troubleshooting	
		Cation conductivity	X		
		Total iron		Once/shift	
		Oxygen	X	Once/day	
		pH	X	Once/shift	
11	Feedwater after economizer	Total iron		Once/shift	
12	Continuous blowdown	Silica		Once/shift (at ≤1000 psig) Twice/shift (at >1000 psig)	
		Specific conductivity	X	Once/shift	
		OH Alkalinity		Troubleshooting	
		M Alkalinity			
		pH	X	Once/shift	
		Product test		Once/shift	To be specified by the water treatment company.

Sample Point No.	Sample Number Description	Parameter	Continuous Analyzer (2)	Minimum Recommended Test Frequency for Manual Testing	Remarks
13	Saturated steam	Sodium (3)	X (4)	Troubleshooting	
		Silica		Troubleshooting	
		Cation conductivity (3)	X (4)		Cation conductivity is preferred to specific conductivity because of its enhanced sensitivity.
		Chloride		Troubleshooting	
		Sulfate		Troubleshooting	
14	Attemperation water	Sodium		Troubleshooting	
		Silica		Troubleshooting	
		Cation conductivity (3)	X (4)		Not needed if run on feedwater and feedwater is attemperation source.
		Chloride		Troubleshooting	
		Sulfate		Troubleshooting	
15	Superheated steam	Same as for sample point 13		Troubleshooting	
16	Desuperheat water	Sodium		Troubleshooting	
		Cation conductivity (3)	X (4)		

(1) Additional tests might be required for higher-pressure boilers or steam generators.
(2) Continuous analyzers should be compared to grab samples at least daily. Increase manual frequency to once/2 hours if analyzer is off line.
(3) Sample point location and sample system are critical.
(4) If required by steam use.

of particulate and metal oxides and prevent contamination during low flow/pressure situations:

- Design sample lines and other system components using inert metal, which is resistant to corrosion by steam and high purity waters (typically 304 or 316 stainless steel).
- Extend the water sample nozzle (tap) into the pipe a distance of 0.23 times the internal radius of the pipe, up to a maximum of 2 inches. The nozzle should be cut at a 45° angle and face into the direction of flow (see Figure 3). A Pitot tube is an acceptable equivalent. If the sample point is to be used for two-phase flow such as for particulate or metal oxides, refer to the section on "Sample Collection."
- Keep sample lines as short as possible, ideally less than 30 feet.
- Minimize the number of valves, fittings and elbows or bends.
- Bend tubing whenever possible rather than installing a fitting.
- Avoid traps and pockets in which fluid or sludge can collect.
- Throttle sample flow at the outlet of the cooler only. Multiple sample coolers may be necessary to achieve the recommended sample temperature.
- Pitch the sample line downward at least 10°, towards the sample outlet.
- Limit sample lines to 1/4 or 3/8 inch tubing to facilitate flushing.

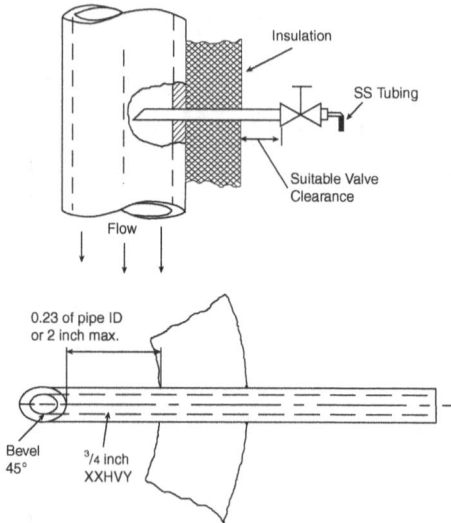

Figure 3 Sampling Nozzle Installation[1]

- Design sample lines in accordance with Table 3 for a velocity of 5-6 ft/sec.
- Design isokinetic sample probes as per ASTM D 1066, *Standard Practice for Sampling Steam* and ASME PTC 19.11, *Water and Steam in the Power Cycle.*
- When sampling superheated steam, locate a roughing cooler as close as possible, but not more than 20 feet from the sample tap in order to "stabilize" contaminant concentrations. Insulate the line between the tap and the cooler.

Table 3 - Purge Times Required for Representative Sampling of Water

Line Size (inches)	Wall Thickness; (inches)	ID (inches)	For Soluble components[1] Recommended Purge Time at 500 ml/min (Sec/ft)	For Particulate Components[2]	
				Recommended Flow Required to Achieve 5 ft/sec (ml/min)	Recommended Flow Required to Achieve 5 ft/sec (gal/min)
1/4 Tubing	0.035	0.180	1.8	1,501	0.40
	0.042	0.166	1.5	1,277	0.34
	0.049	0.152	1.3	1,070	0.28
	0.058	0.134	1.0	832	0.22
	0.065	0.120	0.8	667	0.18
3/8 Tubing	0.035	0.305	5.2	4,310	1.14
	0.042	0.291	4.7	3,924	1.04
	0.049	0.277	4.3	3,555	0.94
	0.058	0.259	3.7	3,108	0.82
	0.065	0.254	3.6	2,989	0.79
1/2 Tubing	0.035	0.430	10.3	8,567	2.26
	0.042	0.416	9.6	8,018	2.12
	0.049	0.402	9.0	7,488	1.98
	0.058	0.384	8.2	6,832	1.81
	0.065	0.370	7.6	6,343	1.68
	0.072	0.356	7.0	5,872	1.55
	0.083	0.334	6.2	5,169	1.37
1/2 Pipe	Schedule 40	0.622	21.5	17,926	4.74
3/4 Pipe	Schedule 40	0.824	37.8	31,459	8.31
1 Pipe	Schedule 40	1.049	61.2	50,985	13.47

(1) Standard, accepted good sampling practices require a minimum of three sample system volumes be purged before a sample is considered "representative" of the system. For example, ten feet of 1/4 inch tubing with a wall thickness of 0.035 inches would require 18 seconds to flush while 100 feet of the same tubing would require 180 seconds or 3 minutes to flush.

(2) High purity water samples typically require longer continuous flushing (24 hours recommended) to achieve the very low concentrations of particulate contaminants present. Flow rate should not be readjusted within 45 minutes of sampling.

5.4 Sampling Frequency

The minimum suggested sampling frequency and monitoring require-ments are shown in Tables 1 and 2. However, if any readings are outside the control limits, the normal sampling frequency must be increased until subsequent readings are consistently within the control limits. Also, during upset conditions, sampling frequency must be temporarily increased until normal control is re-established.

Grab samples are sufficient to provide accurate system information in some instances; however, certain continuous in-line monitors may be required to assure the safety and reliability of the boiler system. These monitors are indicated in Figures 1 and 2. Continuous monitors are preferred.

Collect all grab samples at a frequency that assures the protection and reliability of the boiler system. The recommended/required frequency often depends on the type of internal treatment program used and its ability to tolerate system contamination. Tables 1 and 2 provide recom-mendations on grab sample frequency based on sample location and test parameter.

5.5 Sample Containers and Labeling

Sample containers must be clean and must not interfere with the test and/or preservative procedures to be utilized. The type of capped collec-tion container required is specified by the test method and can be from the following categories:

(a) Borosilicate or Flint glass (either clear or brown)
(b) High density polyethylene plastic
(c) Polycarbonate

Containers are conditioned by performing the following steps:

(1) Clean with mild detergent, rinse and dry
(2) Acid wash
(3) Rinse and soak with high purity water. (High purity water is specified in ASTM Practice D 1193 as Type I)

Refer to Table 4 for sample container type and conditioning required for specific tests.

Clearly label each sample collected with location source, date, and time of sampling.

❏ 5.6 Sample Fixing

Sample fixing may be required for preservation of sample prior to testing, for example iron, copper and TOC. Typically, the pH of the sample is reduced to below 2.0 through the use of one of the following:

(a) Sulfuric acid
(b) Hydrochloric acid
(c) Nitric acid

Maximum preserved holding times (PSHT) prior to testing varies by contaminant. Testing within 15 to 30 minutes of sampling is most preferred.

❏ 5.7 Sample Collection

- Recommended purge times relative to sample line size are discussed in ASTM D 3370, *Standard Practices for Sampling Water* and are shown in Table 3. Sufficient time must be allowed for equilibrium to be established when flushing a sample line or following a change in sample flow rate. High purity water samples typically require longer continuous flushing (24 hours recommended) to achieve representative samples of the very low concentrations of particulate contaminants present. Flow rate should not be readjusted within 45 minutes of sampling.
- Sample flow rate should be constant and flow should be continuous whenever sampling for particulate (e.g. metal oxides in water). Velocities of 5-6 ft/sec. are recommended.
- Use isokinetic sampling probes at their designed flow rate when sampling two-phase systems, such as sodium in steam and iron in water.
- Before taking the sample, rinse the sample container at least three times by filling it to 1/4 of volume with sample, shaking, and then emptying. Refer to Table 4 for the appropriate sample container for collection (normally polycarbonate sample bottles are used except for oil and grease, which normally require glass containers).
- If a preservative or additive is to be used for a specific test, rinse the bottle the prescribed three times, add the preservative, and then collect the sample. Do not overflow the container.
- Samples taken for dissolved oxygen, high purity conductivity and high purity pH must be tested immediately at the sample location or must be determined using a continuous in-line analyzer because of the effect of air absorption on final results. At a sample conductivity of <10 μS/cm, the sample is prone to carbon dioxide absorption from air, which will change both conductivity and pH values.

❐ 5.8 Time Interval Between Sample Collection and Analysis

In general, the time between sample collection and testing should be kept as short as possible. Certain tests results become invalid when samples are stored for any length of time; e.g. oxygen, high purity conductivity, pH, alkalinity, sulfite.

☐ SECTION 6 ☐
ANALYTICAL TESTING

This section provides the user with a general appreciation of the specific analytical tests used for monitoring. Interferences and special considerations to obtaining accurate and meaningful results are provided. The analytical procedures can be found in the ASTM Standards for Water and Environmental Technology, Section II, Volumes 11.01 and 11.02[2]. It should also be noted that other suitable test methods and kits are available commercially. These procedures may be based on the ASTM or Standard Methods for the Examination of Water and Waste Water[3]. The testing may be done on a grab sample (G.S.) basis or continuous sampling (C.S.) utilizing on-line monitors.

There are a multitude of tests, which may be run at the site due to the specific operation. These tests are shown in Table 4, Test Methods. As a note of caution where on-line continuous analyzers are installed, the sample lag time must be determined in order to put the results into phase with real time. This will allow the operator to correlate the data with other events to establish cause and effect relationships. Specialized testing of all internal treatments and organic hydrazine substitutes are beyond the scope of this document. Further information can be obtained from your current water treatment supplier. Other applicable ASTM Standards, which can be referenced, are as follows;

- D 1066 Standard Practices for Sampling Steam (Vol. 11.01)
- D 1192 Standard Specifications for Equipment for Sampling Water and Steam (Vol. 11.01)
- D 1193 Standard Specifications for Reagent Water (Vol. 11.01)
- D 3370 Standard Practices for Sampling Water (Vol. 11.01)
- D 3694 Standard Practices for Preparation of Sample Containers and for Preservation of Organic Constituents (Vol. 11.02)
- D 4453 Standard Practices for Handling Ultra Pure Water Samples (Vol. 11.02)
- D 5540 Standard Practice for Flow Control and Temperature Control for On-line Water Sampling and Analysis (Vol. 11.01)
- ASTM Power Plant Water Analysis Manual[4]

◻ SECTION 7 ◻
TEST METHODS

The particular test method selected will depend upon the accuracy required, particular test equipment available and labor allocation available. One or more of the following analytical techniques are used:

- Gravimetric
- Spectrophotometric
- Colorimetric
- Titrametric
- Ion Chromotography
- Atomic Adsorption or Inductively Coupled Plasma Emission Spectroscopy
- Ion Selective Electrodes (ISE)
- Specialized Vendor Test Kits and Automated Monitors

Commercial test kits are available from numerous sources. The procedures used with these test kits are often based on the ASTM and Standard Method references (cf. Table 4) or modifications thereof.

Attention should be given to the interferences and other considerations listed for each test to be performed prior to sampling. Table 4 provides a list of these. Some general considerations are:

- Calibrate analytical equipment as outlined by the manufacturer.
- Clean and store probes as outlined by the manufacturer.
- Avoid reagent and sample contamination by keeping containers capped; and by working in a well-ventilated dust and fume free area. Avoid reagent contamination by proper handling.
- Adhere to the supplier's reagent shelf life recommendations.
- When preparing standards or reagents on-site, ensure that the reagents used are prepared and standardized as outlined in the specific test method. Adhere to the shelf life constraints in the procedure.
- Continuous sampling for on-line monitors should have the samples pre-cooled to 25°C [77°F], even if temperature compensators are used. The continuous sample flow should be sufficient to ensure the cell chamber is continuously flooded and no dissolved or entrained gases are released. Typical flow range of these prefiltered, constant temperature samples should be in the range of 40 to 500 milliliters per minute.

Table 4 - Test Methods

Test	Reference No.		Type of Method	Detection Range (mg/L)	Interferences and Other Considerations
	ASTM	Standard Methods			
Dissolved Oxygen	D 888	4500-O	Colorimetric (G.S.) Amperometric (C.S.)	0.005–10 0.005–20	Interferences - Color, turbidity, oxidizing impurities and metal oxides may cause higher results with the colorimetric method. Oxidizing gases such as chlorine and nitrogen oxides will provide false positive oxygen values with the probe method. Higher levels of some oxygen scavengers can result in erroneous reading with the colorimetric method. Other Considerations - All sampling lines must be 304 or 316 stainless steel or non-porous synthetic tubing such as neoprene and be leak free. Sample flows should be normalized without adjustments prior to and during testing. Sampling temperatures should be below 30°C [86°F] and held constant. Temperature fluctuations will result in erroneous oxygen concentrations with the probe method.
Total Iron	D 1068	3500-Fe	Colorimetric/ Spectrophoto metric, Atomic Absorption, ICP (G.S.)	0.002–10	Interferences - Copper, cobalt, chromium, nickel, zinc, phosphate, molybdate, and cyanide may interfere with the colorimetric method, if present in excessive concentrations. Strong chelants and other organics may also interfere. Refer to specific method used for details on interferences. Other Considerations - Samples require fixing with hydrochloric or nitric acid. An alternate procedure for suspended iron is to filter a standard volume of water through a 0.45 micron filter. The intensity and color of the stain on the filter is compared with standards to determine the approximate concentration of iron in the sample. Ensure proper sample flow.
Total Copper	D 1688	3500-Cu	Colorimetric/ Spectrophoto metric, Atomic Absorption, ICP (G.S.)	0.002–10	Interferences - Depending on the particular method used, the presence of excess amounts of chromium, tin, cyanide, sulfide, and organic materials may cause interference. Chelating agents interfere at all levels unless removed by digestion. Refer to the specific procedure used for detailed information on interferences. Other Considerations - Samples require fixing with hydrochloric or nitric acid. Ensure proper flow.
Total Hardness	D 1126	2340	Titration (G.S./C.S.) (Atomic Absorption, ICP; Colorimetric)	0.02, minimum	Interferences - The presence of certain metal ions may cause gradual fading or indistinct end points with the titrimetric or colorimetric procedures. Refer to specific procedure used for details on interferences. Other Considerations - If undissolved hardness is to be measured, the sample should be pretreated with acid and then neutralized with sufficient buffer before proceeding with the test. With cold samples, the titration must be done slowly to assure optimum color development.

	ASTM	Standard Methods	Method	Range (mg/L)	
pH	D 1293	4500-H$^+$	Potentiometric (G.S./C.S.)	0–14	Interferences - The pH electrodes are relatively free from interference; however, the response of the electrodes may be affected by temperature and flow. Other Considerations - Follow the manufacturer's instructions for preconditioning of new electrodes and for storage of electrodes. Regularly scheduled calibration, cleaning and maintenance of electrodes is required to assure accurate results. Follow manufacturer's instructions. Calibration of electrodes should be done with buffer solutions having a pH value within one to two units of the sample being tested. There are special challenges when determining pH in high purity water. Use dedicated instruments and test the sample as soon as possible after sampling.
Non Volatile Total Organic Carbon	D 4779 (Test Modification see text)	5310	Infrared detection absorption (G.S.)	0.05–4000	Interferences - Acidification step may lead to partial TOC loss as volatile organic carbon. Other Considerations - Due to potential volatility the sample should be cooled below 37°C [25°C preferred] and the container filled from the bottom. Testing should be done within 24 hours unless refrigerated at 4°C [39.2°F] in an organic vapor free environment. Follow equipment manufacturer's guideline. The method has also been adopted for continuous on-line analysis ASTM D 5173–91. Sampling in brown borosilicate or flint glass containers is recommended. Samples require fixing with sulfuric acid.
Oily Matter	D 3921	5520	Infrared absorption (G.S.) (C.S.)	0.5–100	Interferences - Organic solvents and certain organic compounds not considered as oil and grease (i.e. polymeric dispersants, chelants, antifoams, oxygen scavengers, filming and neutralizing amines) may be extracted and measured as oil and grease. Therefore correction factors may be required. Other Considerations - Sample at 37°C or less [25°C preferred] and fill container from the bottom to avoid volatile losses (standard practice). Sample point should be as close to the extraction point as possible. Calibration curve of known standard concentration and six dilutions are required. All in one analytical instruments are available. Sampling in brown borosilicate or flint glass containers is recommended. Samples require fixing with sulfuric acid.
Sulfite	D 1339	4500-SO$_3^{2-}$	Titration (G.S.) Ion Selective Electrode (C.S.)	1.0, minimum	Interferences - Presence of oxidizable substances such as ferrous iron, organic matter, sulfides, or nitrites will cause falsely high test results. Test as soon as possible, since air absorption may reduce the sulfite concentration. High sample temperatures will prevent proper color formation. Other Considerations - Test as soon as possible. Avoid air contact by filling the container from the bottom, overflowing and then capping.

(Continued)

Table 4 - (Continued)

Test	Reference No.		Type of Method	Detection Range (mg/L)	Interferences and Other Considerations
	ASTM	Standard Methods			
Hydrazine	D 1385		Colorimetric/ (Spectrophoto-metric) (G.S.) Ion Selective Electrode (C.S.)	0.02, minimum	Interferences - Presence of oxidizable substances such as ferrous iron, organic matter, sulfides, or nitrites will cause falsely high test results. High sample temperatures will prevent proper color formation. A minimum of 15 minutes is required for color development. Other Considerations - Samples may be fixed with hydrochloric acid. Test as soon as possible, since air will reduce the concentration. Avoid air contact by filling the container from the bottom, overflowing and then capping. Organic hydrazine substitutes are available. Contact the supplier for applicable testing methods. On-line reductant analyzers are available to determine hydrazine and hydrazine substitute materials concentrations.
Silica, Soluble (Molybdate reactive)	D 859	4500-Si	Colorimetric/ (Spectrophoto-metric) (G.S./ C.S.)	0.005-20	Interferences - Color and turbidity interferences can be removed by dilution or filtration with 0.22 micron filters. Phosphate interferences with color reaction are removed through oxalic acid addition. High dissolved salts (i.e. brines) can affect the color development and can be compensated for through blank and standardization techniques. Strong oxidizing and reducing agents and high organic concentrations can interfere with color formation. Other Considerations - Several steps in the procedure are time sensitive. The color development is pH (1.2-1.5) sensitive. All samples should be collected in new polycarbonate or stainless steel bottles. Glass bottles are not acceptable. See the ASTM procedure for sample container preparation for all other bottles. Silica free reagent and dilution water must be used. This procedure does not detect higher polymeric (colloidal) silica compounds. They can be derived by measuring Total Silica (ASTM D 4517-85) and subtracting the soluble silica value.
Total Alkalinity	D 1067	2320	Titration (G.S.)	1, minimum	Interferences - Natural color and turbidity may interfere when using color indicators. Other Considerations - Results may vary slightly depending on the specific color indicator used. Titrate at room temperature. When measuring P Alkalinity, the sample should be capped and returned to the laboratory for immediate testing to accurately determine the concentration of the various species of alkalinity present in the sample. Immediate testing is not necessary when measuring total alkalinity only.

Test	Reference No.		Type of Method	Detection Range (mg/L)	Interferences and Other Considerations
	ASTM	Standard Methods			
Specific Conductivity	D 1125	2510	Potentiostatic (G.S./C.S.)	$0.05-1 \times 10^6$ µS/cm	Interferences - Metallic solids can short circuit the electrodes. Oils/hydrocarbons, slime, and suspended solids can adhere and foul the electrodes. This leads to a constantly changing cell constant, which hampers balancing the electrical bridge. Gas bubbles adhering to the electrodes also result in erroneous results. Decreasing conductivity can be due to electrode adsorption of filming amines, surfactants or glycolic-based substances. Other Considerations - The test results are very sensitive to temperature, therefore all tests should be run at 25°C [77°F] even if temperature compensated electrodes are used. Mathematical corrections to the standard formula are required if tested above 35°C [95°F]. Continuous sampling units should have sufficient flow to ensure the cell chamber is continuously flooded and no dissolved/entrained gases are released. The sample lines should also be prefiltered and be maintained at a constant temperature. Typical flow rate is 75 to 500 ml/min. For grab samples, testing must be performed as soon as possible after sampling due to absorption of air that results in false readings. The electrodes can foul and the manufacturer's procedures for cleaning should be followed. Grab Sample - Hard glass or chemical resistant containers should be well cleaned and rinsed at least 3 times with the sample water for grab samples. Continuous analysis should utilize stainless steel transport sample lines. Increasing conductivity can be due to atmospheric absorption of acidic gases such as chlorine, carbon dioxide, oxides of sulfur and nitrogen, and hydrogen sulfide, or alkaline gases such as ammonia or organic amines.
Ortho-Phosphate	D 515	4500-P	Colorimetric/ (Spectrophoto metric) (G.S./ C.S.)	0.1-50	Interferences - Sulfite, nitrite and other ions may cause interference. Refer to specific procedure used for appropriate methods of pretreatment to overcome these interferences. Other Considerations - Rate of color formation is time and temperature dependent. Specific color development for individual tests is required.
Total Dissolved Solids	D 5907	2540 C.	Gravimetric (G.S.)	10-20,000	Interferences - Accuracy of results can be affected by sampling methods, sampling system design and conditioning, improper drying, and weighing balance calibration.

(Continued)

Table 4 - (Continued)

Test	Reference No. ASTM	Reference No. Standard Methods	Type of Method	Detection Range (mg/L)	Interferences and Other Considerations
Sodium	D 2791		Ion Selective Electrode (G.S./C.S.)	0.001–1000	Interferences - Interference can occur from lithium, potassium, silver, amine, and ammonium ions at various mg/L levels that produce errors in sodium of about 10%. If the sample pH is above 12, then corrections must be made for the sodium measured at the actual sample temperature. Other Considerations - Sample testing should be done below 40°C [104°F] and should not fluctuate. Lag time response of the sampling and analyzer system must be determined. Sample flow must be constant. If sampling steam, isokinetic sample flow is required as per ASTM D 1066. Grab samples should be taken in polycarbonate plastic bottles.
Total Organic Carbon (TOC)	D 4779 D 5173	5310	Oxidation followed by Infrared Detection Absorption (G.S.) D 5173 for C.S.	0.05–1.0 (D 9779) 0.1–4000 (D 4839)	Interferences - If the TIC (Total Inorganic Carbon) is greater than the TOC, then the TOC should be determined directly by acidification and sparging of the sample. For TIC correction procedure, follow manufacturers recommendation. The process of removing TIC by sparging may result in removal of some volatile organic carbon (VOC) compounds. Acidification may reduce the solubility of a portion of the TOC leading to either syringe pluggage or more VOC loss. High purity waters will readily absorb carbon dioxide; therefore samples should be taken by filling the container from the bottom, overflowing, and then capping with no headspace. This will also minimize loss of volatile organic carbon compounds. Other Considerations - Samples should be taken below 37°C [98.6°F], preferably 25°C [77°F] to avoid volatility loss. Samples to be tested within 24 hours unless refrigerated at 4°C [39.2°F] in an organic vapor free environment. Fixing of samples and sample containers with phosphoric acid to pH 2 is required. Sample containers should be TFE-Fluorocarbon lined, brown Borosilicate, or Flint glass. Sample lines should be TFE-Fluorocarbon lined or stainless steel. Sample flows should be at 500 ml/minute or more; do not readjust sample lines after purging and prior to or during sampling. Rinse sample containers three times with the sample while filling from the bottom of the container. It is very important that the sample bottles are pre-cleaned prior to use as per the ASTM method.

Test	Reference No.		Type of Method	Detection Range (mg/L)	Interferences and Other Considerations
	ASTM	Standard Methods			
Cation Conductivity	D 4519		Potentiostatic, electrical specific conductivity that utilizes a hydrogen form cation resin column to remove all cations and convert anions to the corresponding acids. (G.S./C.S.)	0.05–10 µS/cm	Interferences - Metallic solids or ion exchange resin beads can short-circuit the electrodes. Oils/ hydrocarbons, slime and suspended solids can adhere and foul the electrodes. This leads to a constantly changing cell constant that hampers balancing the electrical bridge. Gas bubbles adhering to the electrode also result in erroneous results. Other Considerations - Cation column must be conditioned using high purity water (D 1193 Type II specifications) to achieve baseline high purity values. Spent resins will also affect the results since no ion exchange occurs. Flow rate through the column can affect the results as re-exchange can occur. The manufacturer will provide the optimum flow range, which is typically 200 to 500 ml/min on a 1 5/8 to 2 inch inside diameter column. The results are affected by temperature; hence the samples should be cooled to 20–25°C [68–77°F]. Will detect carbon dioxide, inorganic and organic anions including organic acids. Color changing cation resin columns with see through chambers are available to indicate when resin is near exhaustion.
Degassed Cation Conductivity	D 4519		Potentiostatic, cation conductivity that also utilizes an inert gas sparging or a reboiler/ chiller to remove carbon dioxide and low molecular weight, volatile, purgable organic acids	0.05–10 µS/cm	Interferences - See Specific and Cation Conductivity. Other Considerations - See Specific and Cation Conductivity. Tables and graphs are provided by the manufacturer of the equipment for interpretation of the results if the microprocessor option is not present. Continuous on-line analysis uses an atmospheric reboiler chamber.

(Continued)

Table 4 - (Continued)

Test	Reference No.		Type of Method	Detection Range (mg/L)	Interferences and Other Considerations
	ASTM	Standard Methods			
Chloride/ Sulfates	D 5542	4500-Cl 4500-SO$_4$$^{-2}$ EPA 300	Ion Chromatography (G.S. D 5542, C.S. D 5996)	0.002–0.1	Interferences - Samples containing high concentrations of certain anions may cause very large peaks on the chromatogram that could mask other anion peaks that are present. Normally, dilution can minimize this effect. High concentrations of certain anions may cause low breakthrough volumes of other anions from the concentrator column. Do not attempt to concentrate a volume of sample greater than 80% of the breakthrough volume. Other Considerations - Samples containing high (mg/L) concentrations of ammonia, amines, or other additives that raise the hydroxide concentration (pH) of the sample may cause low break through volumes. This can be avoided by taking the sample after the cation resin of a cation conductivity detector. Cation suppressor membrane can foul.
ORP (Oxidation Reduction Potential)	D 1498		Potentiometric (G.S./C.S.) Measures the net potential of the solution to oxidize or reduce species in the solution	–1000 to +1000 mV	Interferences - The electrodes are relatively free from interference; however, the electrode response is affected by temperature and pH. As the temperature increases or the pH decreases, the measured ORP will increase. Other Considerations - Follow the manufacturer's instructions for preconditioning of new electrodes and for storage of electrodes. Regularly scheduled calibration (i.e. within 30 mV of expected standard), cleaning and maintenance (i.e. successive readings should not vary more than ±10 mV) of electrodes is required to assure accurate results. Brightly polished platinum or gold electrodes are required to avoid the potential for adsorption of materials that may affect the accuracy. Shielded cable for signal transmission is required for remote measurements.

❒ BIBLIOGRAPHY ❒

1. Allmon, W. E. and S. J. Potterton, "Chemical Analysis for Fossil Plant Water Chemistry," EPRI Symposium on Fossil Plant Water Chemistry, Atlanta, Georgia, June 1985
2. Annual Book of ASTM Standards Volumes 11.01 and 11.02, American Society of Testing and Materials, 1916 Race Street, Philadelphia, PA 19103-1198 USA, ISBN 0-8031-2184-9
3. Standard Methods for the Examination of Water and Wastewater, prepared and published jointly by the American Public Health Association, American Water Works Association, and Water Environment Federation, Publication Office: American Public Health Association, 1015 Fifteenth Street, N.W., Washington, DC 20005, ISBN 0-87553-207-1
4. ASTM Power Plant Water Analysis Manual, American Society of Testing and Materials, 1916 Race Street, Philadelphia, PA, 19103, ASTM Publication Code No. (PCN) 03-419084-16, ISBN 0-8031-0200-3
5. "Consensus on Operating Practices for the Control of Feedwater and Boiler Water Chemistry in Modern Industrial Boilers." Water Technology Subcommittee of the ASME Research and Technology Committee on Water and Steam in Thermal Systems, ASME CRTD Vol. 34
6. "Consensus Guideline on Fossil Plant Cycle Chemistry," EPRI CS 4829, Project 2712
7. Groose, J. E. et al, "Trends in Water Sampling Technology for Evaluating Corrosion in Steam Generating Systems," National Association of Corrosion Engineers Meeting, March 1993
8. "A Practical Guide to Avoiding Steam Purity Problems in the Industrial Plant," Water Technology Subcommittee of the ASME Research and Technology Committee on Water and Steam in Thermal Systems, ASME CRTD Vol. 35
9. "Consensus for the Lay-Up of Boilers, Turbines, Turbine Condensers and Auxiliary Equipment," Water Technology Subcommittee of the ASME Research and Technology Committee on Water and Steam in Thermal Systems, ASME CRTD Vol. 66
10. "Consensus on Operating Practices for Control of Water and Steam Chemistry in Combined Cycle and Cogeneration Power Plants." Water Technology Subcommittee of the ASME Research and Technology Committee on Water and Steam in Thermal Systems